THE BOOK OF FORMULAS

JOHN HAZELRIGG, F.H.S.

Philip N. Wheeler, Editor

Alembic Publishing LLC

John Hazelrigg

Published in 2011 by Alembic Publishing LLC
P. O. Box 1523
Waynesboro VA 22980-1392
http://www.alembicpublishing.com

First Edition 2011 by Alembic Publishing LLC

ISBN: 1468103636
ISBN-13: 978-1468103632

Cover design and computer graphics by Philip N. Wheeler based on illustrations in ancient alchemical texts.

This book is sold for informational purposes. The author and publisher will not be held accountable for the use or misuse of the information in this book.

Printed in the United States of America

A COLLECTION OF CHOICE SPAGYRIC PREPARATIONS
PREDESTINATED TO THE CURE OF DISEASE AND
THE CONSERVATION OF HEALTH: INCLUDING
SOME OF THE RAREST AND MOST
VALUABLE SECRETS OF THE ANCIENT
MEDICAL AND HERMETIC
PHILOSOPHY.

COLLATED AND RENDERED INTELLIGIBLE
WITH EXPLICATORY ANNOTATIONS.

John Hazelrigg

TABLE OF CONTENTS

Disclaimer

This document contains historical information translated from texts written over 100 years ago. The information herein is presented for informational, historical and research purposes. Some of the processes described involve extremely poisonous materials that should only be used by persons who are professionally trained in safe materials handling and environmentally approved disposal procedures.

Liability: The publisher does not warrant or assume any legal liability or responsibility for the accuracy, completeness, or usefulness of any information, apparatus, product, or process disclosed. The publisher makes no representation as to the accuracy or completeness of the contents of this book and specifically disclaims any implied warranty of merchantability or fitness for a particular purpose. No warranty may be created or extended by written sales materials or sales representatives. You should obtain professional consultation where appropriate. The publisher shall not be liable for any loss of profit or other commercial or personal damages, including but not limited to special, incidental, consequential, or other damage.

To The Reader

Philip N. Wheeler

The author of the original work, John Hazelrigg (1860-1944), was an American astrologer. He co-founded the American Academy of Astrologians in 1916. Mr. Hazelrigg took an interest in alchemy and in particular the processes related in the writings of various alchemists of the past several centuries. Mr. Hazelrigg's publications include the following:

Metaphysical Astrology. New York: Metaphysical Publishing, 1900.

The Book of Formulas. New York: Hermetic Publishing, 1904.

The Sun Book, or the Philosopher's Vade Mecum. New York: Hermetic Publishing, 1916.

Astrosophic Principles. New York: Hermetic Publishing, 1917.

Yearbook of the American Academy of Astrologians. New York: Hermetic Publishing, 1917.

Fundamentals of Hermetic Science: Being the Bona Fides of Astrology. New York: Hermetic Publishing, 1925.

Hans Nintzel released a version of The Book of Formulas in 1977 as part of the R.A.M.S.[1] collection of alchemical works. In addition, The Book of Formulas has been widely available for many years in print and electronic form. As a tribute to my friend Hans, I have prepared this annotated edition for modern researchers, historians and students of the art of alchemy.

The symbols originally in the 1904 edition are redrawn in high-resolution and presented with their definitions in the Symbol

---

[1] The Restorers of Alchemical Manuscripts Society (R.A.M.S.) was organized by Hans Nintzel (d. 2000) in the 1970s.

Dictionary. Astrologers traditionally use symbols for the signs of the zodiac, celestial bodies, aspects, and lunar phases. Many of those symbols carry over in the study of alchemy, usually with an added meaning. A much more complete presentation of alchemical symbols may be found in *Tables of Alchemical Symbols and Gematria,* listed in the References section. It contains over 450 alchemical symbols and other valuable alchemical reference material.

This edition includes edited text referenced from a copy of the original 1904 edition. Unusual, possibly erroneous spelling in the original text is included as a footnote. The goal is to be true to the original text.

The text itself contains my English translations of the symbols. Within the text a superscript such as "s1" indicates an entry in the Symbol Dictionary, whereas a simple numerical superscript indicates a page footnote. For example, "water s4" refers to a symbol from the original text, whereas "water" without the "s4" superscript is how the word or phrase appeared in the 1904 edition.

In several cases there are suspected typesetter errors in the original edition. For example, sulfur s29 is first seen as sulfur s27, so noted and with an added footnote of explanation. I have never before seen sulfur abbreviated as shown by the symbol in "s27" and I suspect that the typesetter inverted the symbol.

Annotations were added by the current editor regarding various chemical names and molecular formulas as well as explanations of many alchemical and medical terms. These annotations appear as numbered footnotes. The "Notes" in the primary text are those of Hazelrigg.

It is my hope that these efforts contribute in some small way to future alchemical research, toward which my friend Hans Nintzel devoted many years of his life.

PREFACE
John Hazelrigg

I take great pleasure in here presenting to the earnest few of beneficent and uplifting purpose this collection of valuable forms and medicaments, culled and transcribed into more lucid phrase from the chemical and Hermetic writings of such learned worthies as Paracelsus[2], Faber[3], Quercetan[4], Hartmann[5], Rulandus[6] and Mynsicht[7], and numerous other philosophers in the Spagyric Art. That these preparations -- and many similar ones which I hope to treat of in a subsequent work -- have remained absent from the

---

[2] Paracelsus (Philippus Aureolus Theophrastus Bombastus von Hohenheim, 1493-1541) was a Swiss Renaissance physician, botanist, alchemist, astrologer, and general occultist.

[3] Faber: Possibly Jakob Faber, an alchemist working in Bohemia in 1585.

[4] Quercetan: Joseph Duchesne or du Chesne (Quercetan, Latin Josephus Quercetanus) (c.1544-1609) was a French physician. A follower of Paracelsus, he is now remembered for important if transitional alchemical theories.

[5] Franz Hartmann (1838-1912) was a German physician, theosophist, occultist, geomancer, astrologer, and author.

[6] Martin Ruland the Elder (1532-1602), also known as Martinus Rulandus or Martin Rulandt, was a German physician and alchemist. He was a follower of the physician Paracelsus. His son, Martin Ruland the Younger (1569–1611), also became a renowned physician and alchemist.

[7] Adrian von Mynsicht (1603–1638) was a German alchemist. He is best known for the allegorical work <u>Aureum Saeculum Redivivum</u>, published under the pseudonym Henricus Madathanus, and usually dated to 1621. It was soon reprinted in the collections <u>Musaeum Hermeticum</u> and the <u>Dyas Chymica Tripartita</u>, both in 1625 (Frankfurt).

officinal[8] category is not a matter for wonder, chiefly because of a recognition by their authors of specific powers and potencies in natural processes of too arcane a nature to be apprehended by the dense self-sufficiency of the materialistic and incredulous mind; and partly by reason of the fact that while dealing with obvious principles, though only occultly understood, they were couched in terms and meanings of which was wisely veiled from those of disrespectful and antagonistic attitude.

The importance to the occult chemist of working in harmony with astral law, that thereby the natural and artificial qualities of all things from the Universal to the Particular might be properly attained, may be the better appreciated in this extract from Salmon[9], an astro-philosopher and physician of the Middle Ages:

"1. The time of the preparation ought to sympathize with the native production of the thing to be prepared; which is in respect of qualities manifest or occult.

"2. As to the Manifest Qualities, that time is to be chosen in which they naturally flourish; wherein you are to choose a hot and moist season for dissolution, digestion and fermentation; and a cold time for coagulation; a moist time for distillation and melting; and a dry time for exsiccation and calcining.

"3. As to the Occult Qualities, the preparation is to be begun when the planet governing the thing is strong and vigorous in his

_____

[8] *Officinal*, not to be confused with the word "official," is a term applied in medicine to preparations that are sold in a shop that formulates and dispenses materia medica to physicians, surgeons and patients.
[9] William Salmon (1644-1713), an alchemist, author and translator of numerous alchemical works. Note that, Hazelrigg's descriptor notwithstanding, this period is not the Middle Ages; it is the late Renaissance/early Enlightenment.

house or exaltation, and in good aspect of Sol, Luna, Jupiter, Venus, or all of them.

"4. The place of preparation must be the laboratory, which must be hot, cold, moist, dry, airy, close, etc., according as the nature of the matter to be prepared requires."

The significance of planetary influence, and its relation to the astral potencies involved in all natural operations, is too complex a subject to enter into here, except to say that the truth of the above brief intimations has been amply verified in the writer's experience. Nor when the rationality thereof is once understood, through careful study and investigation of the stellar hypothesis, will one marvel that it should be so.

Likewise, in dealing with spiritual principles of things physical, the Spagyric artist was enabled to perceive the admirable analogies that helped verify the oneness of method throughout the spheres of manifestation, and its identity with those of the Higher or Causative realm, and thereby the necessary interaction and dependence of the one upon the other. Proceeding thus, under the logical assumption that the One Law must express itself similarly upon all planes of activity, they demonstrated the reality of a physical trinity -- spagyrically classified as Salt, Sulphur and Mercury -- that corresponded with the Body, Soul and Spirit of the nominal world, or the Father, Son and Holy Ghost of the devotional school; also the fact that these three primal principles embrace and comprehend the four elements, Earth, Fire, Air, and Water, the separation, purification and inseparable conjunction of which constituted a fifth, of the purest potency, which they termed a Quintessence. This on the spiritual plane, as embodied in the esoteric teachings of the ancient religious mystics, is identical with Regeneration, a process the meaning of which modern churchianity knows as little of as the material scientist does of the above trinity in nature. With this attainment in chemical processes all poison has been eliminated from the matter operated upon, and the spiritual or curative faculty is exalted to the highest degree, as is shown in some of the formulas which make up this collection. The antimonial prescripts of Basil Valentine[10] and the mercurial preparations of Paracelsus were not those which pass today as medicines, but which in reality are poisons most dangerous to the vital principle in the physical organism. The alchemists scorned the use of remedies that yet remained in such imperfect guise.

In a later treatise I hope to elaborate more fully and practically upon the Hermetic Philosophy as concerns the preparation of metallic, mineral and vegetable medicaments, their spiritual bases, and whereby their different faculties may be developed into the highest vital energies.

With these concise hints by way of introduction, I commend the contents of the following pages to those of earnest mind, and particularly to such whose object is the mitigation of human physical suffering -- one of the grandest missions to which an earthly pilgrimage can be consecrated.

---

[10] Basil Valentine is the author of *The Triumphal Chariot of Antimony* and other well known alchemical works. Basilius Valentinus was allegedly a 15th-century alchemist. There are claims that he was the Canon of the Benedictine Priory of Sankt Peter in Erfurt, Germany but according to John Maxson Stillman there is no evidence of such a name on the rolls in Germany or Rome and no mention of this name before 1600. During the 18th century it was suggested that the author of the works attributed to him was Johann Thölde.

MENSTRUUMS
John Hazelrigg

According to Dr. Johnson[11], the use of the word *menstruum*[12] originated in the notion of the old chemists respecting the influence of the Moon in their preparations -- a fact that contains the essence of a truth such as the modern chemist, unfamiliar with the astral principia, are both indisposed to concede and unable to comprehend.

In all the formulae contained in this book, the practitioner is advised to use only such menstrua or dissolvents as are made in the private laboratory, for those of commerce are undependable, generally by reason of the impurity of the substances used, or else through an utter disregard of the spiritual principles involved. A perfect exaltation of medicines can never in any wise be accomplished through adulterated or imperfect mediums.

Divers chemical experiments delivered by the ancients have been believed false, only because the menstruums employed in the highly unsuccessful trials of them were not as highly rectified, or otherwise exquisitely depurated[13], as those that were used by the deliverers of those experiments; so that oftentimes the fault of a bad menstruum is injuriously imputed to a good artist. Many such purchased in the apothecary shop are wholly unfit; barely by their

---

[11] Samuel Johnson was an English author of the 18th century.
[12] Traditionally, a relationship between the human menstrual cycle and the lunar cycle has been posited.
[13] Cleansed or purified

not being sufficiently freed from their weakening aquosity[14], as is very often the case with spirit of wine. On the other hand, while some processes fail to succeed according to the expectation because the menstruums employed about them were not pure enough, so some miscarry because such menstruums are but too exactly depurated. Again, while the purity of these is a prime desideratum, one principal regard is as to their fitness for the particular purpose to which they are designed. Thus, an aqua fortis that is proper in one operation must be differently complexionated for the better success of another. For which reason the different solvents, as here given for use in the succeeding formulas, are successively numbered, and so referred to therein in brackets[15], that the essential one may be utilized.

In most instances of digestion and putrefaction, and more especially where a quintessence is the end to be attained, a homogene menstruum — as the spirits, phlegm or water of the subject matter itself — is absolutely a requisite, that the astral principles inherent therein be not fatally disturbed, or their efficacy impaired. As saith Paracelsus — "Every fruit must die in that wherein is its life" (Archidoxes[16])

Modern chemists, for the sake of expediency and convenience, and largely through ignorance of Nature's interior methods, operate

---

[14] The condition of being wet or watery.

[15] The author includes these as, for example, (10a) in the original text. These menstruums are explained in the next section. They are referenced in many of the recipes presented in subsequent sections. To avoid confusion I have added the word "Menstruum" to each occurrence. Thus, "(1)" in the original text is "(Menstruum 1)" in this edition.

[16] From *Archidoxes of Magic* by Theophrastus Paracelsus (1493-1541). Seeds are the key to regeneration and the continuance of life for that species.

with mediums foreign to their matter, and thereby injure its natural or curative potencies.

I give here careful and complete directions for the making of such menstruums as are necessary in the medicinal formulas that follow.

TO PURIFY MINED NITRE[s1]

The ordinary way of purifying all salt is by simple solution, filtration and crystallization. But note this in the purification of all mined nitre[s1] [17], that after solution and filtration you must digest[s2] for a medical month (40 days), beginning at the wane of the Moon, when putrefactive processes are more easily carried forward, then evaporate and crystallize. This digestion[s2] is necessary before the solution can properly be separated from the fæces. Repeat the operation till no more fæces will settle. According to Paracelsus, dissolve[s3] it in water[s4], filter, and evaporate, till a crust appears at the top, place in a cool place to crystallize, evaporate the water[s4] again, and again set to crystallize, which repeat thrice more.

Common salt[s5] and nitre[s6] are purified by solution in ordinary water[s4] followed by filtration and crystallization, without the digestion[s2].

---

[17] Sal Gemma: salt adrom, sendaro sabachi, metalline salt, Hungarian salt, sparkling salt - *Lexicon of Alchemy* By Martinus Rulandus. I read the symbol as "mined niter" or "unrefined potassium nitrate from the mine."

COMMON AF. (AF. COMMUNIS, AQUA FORTIS[18])

(Menstruum 1)

Rx.[19] Dried Hungarian mined niter[s1] (prepared as for oil[s7] of mined nitre[s1]) 2 pounds[20], mix and put into a retort two-thirds full, with a large receiver, place in a little reverberatory[21] fire[s8]. Proceed with the first degree of fire[s8] till the phlegm disappears; when the red vapors fill the receiver, then increase the fire[s8] by degrees till the spirit is drawn off. This should take twelve hours or longer. It is known by its yellowness, or greenish color, and its stinking smell. This is a dissolvent for silver[s9] and copper[s10].

NOTE: The HNO_3 of commerce lacks the essential disintegrative potencies of the above, and will be found of doubtful utility for the purposes needed. Out of nitre[s6] and mined nitre[s1] undephlegmated[22] 1 pound you may have 16 ounces[s11] of aqua fortis. For certain other operations, AF. (aqua fortis) is made of nitre[s6] p.i. (one part nitre) ammonium chloride[s13] p.i. (one part ammonium chloride).

---

[18] AF.: Aqua fortis, nitric acid, or spirit of niter. Molecular formula HNO_3.

[19] Rx is a symbol for medical prescription.

[20] In the original text this appears as "lbii." where "lb" = pound, and "ii" = Roman numeral representing 2. The rest of this text follows this convention.

[21] A reverberatory furnace is a furnace that isolates the material being processed from contact with the fuel, but not from contact with combustion gases.

[22] Undephlegmated: Not deprived of superabundant water, as by evaporation or distillation.

AR. (Regis communis, AQUA REGIA[23])

This passes in the modern dispensatory as nitromuriatic acid, and as chloro-nitrous acid (NCl_2O_3), this latter being merely a mixture of chlorine and hyponitrous acid. While it dissolves gold, platinum, etc.[24], it does not extract the spiritual tincture. Like many other chemical preparations of commerce, it fails in the purpose for which it was designed. A common aqua regia that will dissolve gold[s15] is thus prepared:

(Menstruum 2)

Rx. 1 pound of aqua fortis, common salt[s5] decripitated[25] 4 ounces[s11], distil[s16] with a strong fire[s8] three or four times till all the salt[s5] comes over with the spirit; in every 3 ounces[s11] of which solution[s3] ammonium chloride[s13] [26], thrice sublimed 1 ounce[s11]. Or thus: Rx. spirit of nitre[s6] 3 ounces[s11] ammonium chloride[s13] 1 ounce[s11], digest. Or,

---

[23] AR. Royal water. Composed of nitric acid and hydrochloric acid, usually mixed in a 1:3 volume ratio.

[24] Aqua regia dissolves the so-called noble metals gold and platinum. It does not dissolve ruthenium, tantalum, iridium, osmium, titanium, rhodium and a few other metals.

[25] Decripitate: to heat a substance, such as a salt, until it emits a crackling sound.

[26] NH_4Cl

(Menstruum 3) Aqua Regia of Tentzelii, The Golden Dissolvent of Tentzelius[27].

Rx. Nitre[s6] 1 pound, powder of flints 3 pounds, distil the water from them by a retort; from the caput mortuum extract the nitre[s6] with hot water; with the remaining ashes mix ammonium chloride[s13] 1 pound, distil by a retort in a naked fire, so have you aqua regia; which distilling in balneum maria[28] or ashes in a glass alembic, will yet be clearer. Or,

(Menstruum 4)

Rx. Dissolve ammonium chloride[s13] 6 drams[s14] in aqua fortis 1 pound. Or, according to Basil Valentine, the alchemist of the XIVth century,

(Menstruum 5) AR. Basilii, Basil's Dissolvent of Gold.

Rx. nitre[s6], ammonium chloride[s13] 2 pounds each, powder of flints 3 pounds, distil by a retort red hot with a pipe. Schroder[29] saith the ordinary way is by an iron pot, on which is placed an earthen cover like an alembic with a beak, set into a naked fire, with a large

---

[27] Andreas Tentzelius was a 17th century European author who wrote at least seven books on alchemy.

[28] BM. A bain-marie (French term for a water bath) is a piece of equipment used in science, industry, and cooking to heat materials gently and gradually to fixed temperatures. It resembles a double boiler.

[29] Friedrich Joseph Wilhelm Schröder (1733-1778) was a physician, alchemist and Rosicrucian. From 1764 he was a professor at Marburg University.

receiver. This not only dissolves gold[s15] , but carries the dissolved gold with it through the receiver.

OF SAL NITRE[s6]

The ancient chemists denominated this under various names, viz., Cerberus, Salt of Hermes, Anderonae, Anatron, Cahalatar, Infernal Salt; Basil called it the Earthly Serpent. That of commerce comes principally from Chili and Peru, and is often adulterated by mixture with common salt[s5], a fact that can be detected by burning, for being fired upon a red hot tile or stone if it all fly away it is pure, however if anything remain it is common salt[s5]. It is purified by solution, filtration and crystallization in common water according to the usual way.

(Menstruum 6) Spirit of Niter[s6]

Rx. Pure nitre[s6] 1 pound. Potters earth 5 parts, mix them, with which fill up to the neck a glass retort, well luted up to its neck, place it in a close reverberatory furnace, with a capacious receiver; give it fire by degrees to the highest for the space of 24 hours; the phlegm comes first with white vapors, then the spirit in red vapors, which will make the receiver as red as a ruby.

NOTE: See that the receiver shall be well and close luted to the retort, lest the vapors should exhale, as their malignancy is very hurtful to the brain and nerves, tending to paralysis. The phlegm that is mixed with the spirit will do it no injury, because it will be so small a proportion to the spirit as to make it lose none of its energy or force.

(Menstruum 7) Bezoardic[30] Spirit of Nitre[s6]

Rx. Rectified spirit of nitre[s6], butter of antimony[s19] (this is antimony trichloride, $SbCl_3$), in equal parts[31], distil[s16] in a retort, so have you the Bezoardic Spirit of nitre[s6]. The caput mortuum serves to make Bezoar Mineral of nitre[s6]. If the above be strongly forced the spirit will be red, and be able not only to dissolve gold[s15], but also to volatilize it.

The Bezoardic Spirit of Common nitre[s6] is made similarly to the above.

---

[30] A bezoar is a mass found trapped in the gastrointestinal system. Bezoars were sought because they were believed to have the power of a universal antidote against any poison.

[31] In the original text this was "ana," which means "in equal parts."

OF COMMON SALT[s5]

This is purified by solution and crystallization till it be without fæces and sweet. Salmon says that salt[s5] should first be decrepitated, lest being put into any hot vessel it should break it. And Basil calcined common salt[s5] thrice with lime, mixing with it every time as much fresh quicklime.

(Menstruum 8) Common Oil or Spirit of Salt[s5].

Rx. Of French or Spanish Bay salt[s5] as much as you will, dissolve it in water and filter it; mix with this brine (in a copper vessel) fine powder of tiles or brick, double or treble the weight of the salt[s5] before dissolution, set it in an equable heat and let the water evaporate away (continually stirring it) till it be very dry; then put the powder in a glass retort, well luted to a large receiver, and give it a naked fire by degrees to the height, so shall you have in the receiver Oil or sharp Spirit of Salt[s5]; rectify this liquor in sand by separating the phlegm, according to our art, then keep it in a vessel close stopped for use.

NOTE: There are a variety of ways of preparing this oil or spirit, according to the opinions of various men. Some take common salt[s5] and distil it alone in a retort, whence comes first a sour water, secondly a phlegm, then the caput mortuum[s18] being taken out and sprinkled with water and distilled again, you have thirdly an acid spirit, which you may repeat till all the salt[s5] is turned into spirit, which rectify in BM. Some to salt[s5] 2 parts, take of clay, brick, sand or earth, 3 parts, which they make into balls and dry and distil as before in a retort. Others to salt[s5] 1 part mix clay, sand, earth, etc., 5 parts and (without making it into balls) fill a glass retort with it, luted up to half the neck, then set it into a close reverberatory, fitting to it a large receiver, thence giving fire by degrees, coming as soon as possible to the last degree, which is to be continued 24 hours or until the recipient feels cold, the retort being violently hot. By this rule you may extract out of 1 pound of salt[s5] near 10 ounces[s11] of oil or spirit. In these processes the salt[s5] ought not to be decrepitated or separated from its phlegm, or the matter made red hot in the fire; for the phlegm coming first helps the acid oil or spirit, and is as a vehicle to it, and without which you would attain to no acid spirits. This spirit is made strong and pure by the dephlegmating of it in a sand heat.

This Oil or Spirit of Salt[s5] is an antidote against the plague, and resists all manner of corruption, both inwardly taken and outwardly applied; it is a specific against malign fevers, whether continued or intermitting; mixed with a little water it whitens and preserves the teeth wonderfully; it is a powerful diuretic, and the most formidable enemy to the scurvy; it opens all obstructions of the stomach, liver, spleen, bowels, reins and bladder, thereby dissolving the stone and gravel and expelling of it; it is good in jaundice, gouts, etc.; it quenches thirst wonderfully, and consumes by corrosion without pain all corruptions in ulcers or other old sores, fistulas, etc. It is transparent and almost of a citrine color, and is of an acid taste. Dose 6 to 15 drops[32] or 20.

---

[32] The original text read, "*à gut* vi. *ad* xv." that means "6 to 15 drops." The rest of this text follows this convention commonly used in apothecary measures.

(Menstruum 9)

Rx. volatile salt[s5] of urine[s20] [33](1 pound), common salt[s5], ammonium chloride[s13], and potassium bitartrate[s21], in equal parts, mix and put to them spirit of vinegar[s22], digest in a luted vessel for an hour, then in ashes distil to dryness; to these salts thus dried 10 pounds add tripoli[34] 30 pounds, and draw a spirit by a retort with a strong fire. This is used as a menstruum in the making of the mercury[s23] of antimony[s19] that enters into the formula for Tinctura Antimonii cum Auro.

NOTE: There are many other preparations of salt[s5], of rare value as medicines, and by which gold is not only dissolved, but volatilized and made potable. The hydrochloric acid (HCl) of commerce is of no utility in these processes.

---

[33] Probably ammonium carbonate.
[34] Tripoli is a naturally-occurring, special application abrasive mineral used in a variety of industries for sharpening, buffing and polishing.

OF POTASSIUM BITARTRATE[s21]

(Menstruum l0a) Spirit and Oil of Potassium Bitartrate[s21]

Rx. Powder of white potassium bitartrate[s21], distil it with a glass retort in sand (or a naked fire); first you have a phlegm, then a spirit like a cloud, and lastly a thick oil, which separate from the spirit. The spirit you may rectify by three cohobations[35] upon colcothar[36], or by distilling it four times in a BM., always washing the still with a strong lixivium[37]. The oil is rectified by adding to it a good quantity of water or distilled vinegar, and distilling in BM. Separate the oil, and mix with it again good rose-water; distil, separate and keep the oil for use.

NOTE: The spirit is sudorific, diuretic, antiscorbutic and anodyne[38]. It cuts, attenuates, dissolves, and opens all obstructions and is wonderful in dropsies, gout, scurvy, palsy, scabs, itch, leprosy, or French-Pox. Dose, 20 drops to 2 scruples[s12] [39].

---

[35] Cohobation is the process of repeated distillation of the same matter, with the liquid drawn from it; that liquid is poured again and again upon the matter left at the bottom of the vessel.

[36] A brownish-red oxide of iron, obtained by heating ferrous sulfate.

[37] A solution, containing alkaline salts, obtained by leaching wood ashes with water; lye.

[38] A pain-killing drug or medicine.

[39] A scruple is an old apothecary measure equal to 1/3 dram. There are 8 drams in an ounce.

(Menstruum 10b) Philosophic Spirit of Potassium Bitartrate[s21]

Rx. Salt[s5] of potassium bitartrate[s21], which reverberate 24 hours upon a refiner's test (but melt it not), and the salt[s5] will be blue; add distilled vinegar[s22] three inches above it, digest[s2] till the vinegar[s22] is very red, then filter and coagulate; do this four times with fresh distilled vinegar[s22]. Mix these four salts with rectified SV.[40], and extract a tincture till it ceases to be colored, dissolve[s3] the salt[s5] remaining in distilled vinegar[s22], extract its tincture with SV., gather all these and distil[s16] them in BM. in a still with an alembic and a funnel; pour in again the SV. distilled at the funnel to the tinctured salt[s5], and cohobate 15 or 16 times, till you see red drops fall; then cease and distil[s16] the liquor gently by a retort in sand. So have you the spirit of salt[s5] of potassium bitartrate[s21], cloudy with red drops and a red powder at the bottom.

This spirit cures quartans and powerfully provokes the Terms being stopped, at the third or fourth time it is taken. Dose, 4 drops or 6 in wine. This spirit will dissolve gold[s15].

(Menstruum 11) Oleum potassium bitartrate[s21] per deliquium[41].

Rx. Let salt[s5] of potassium bitartrate[s21] melt in a cellar or dissolve it in water, filter, and coagulate to the just consistency[42]. Or, salt[s5] of potassium bitartrate[s21], or best of potassium bitartrate[s21] calcined

---

[40] Spiritus Vini, or Spirit of Wine: A rectified solution of ethyl alcohol prepared by distilling wine. "SV." is used throughout the original text.
[41] A melting or liquefaction by absorption of moisture, as of a salt.
[42] The author's meaning is unclear.

white; put it into a cotton bag, and hang it in a cellar or in some moist place to dissolve, and then filtrate.

NOTE: A little of potassium bitartrate[21] mixed with any menstruum facilitates putrefaction, and makes it extract the virtues of any vegetable the easier.

John Hazelrigg

OF MINED NITRE[s1]

Attention is here called to foregoing remarks upon mined nitre[s1], as also what is said concerning its purification. Mined nitre[s1] is either native or factitious, the former being taken out of the earth either in its own form or in water. The factitious is made out of copper[s10] or iron[s24], severally or conjunctly. Of these kinds it will be good to choose that which has more copper[s10] than iron[s24] in it, the Hungarian being the best, which rubbed upon bright iron makes it look red.

(Menstruum 12) Oil of Mined Nitre[s1]

Rx. As much of native or Hungarian or English mined nitre[s1] as you please; melt it in an unglazed earthen pan, and exhale away all the humidity, continually stirring it till it is brought into a yellow powder, which place in a retort that will endure a strong fire, filling it about two-thirds full. Place on an open fire, which give by degrees for three days, or until the receiver (having been full of fumes) becomes clear, and the spirit or oil comes; rectify the distilled liquor, separating the phlegm by a small retort in sand. Note, in distilling, the phlegm comes first by a very small fire, then increase by degrees to the highest, which continue till you perceive black veins trickle down the recipient; which then remove, decant the phlegm, and fit the receiver again without luting, to take the oil. The phlegm being separated in distilling, what comes after it is spirit and oil; separate in a glass cucurbit luted to a receiver, drawing off about one half-part, which is the volatile and sulphureous spirit of mined nitre[s1], which keep in a vial close stopped; what remains at the bottom is the caustic oil of mined nitre[s1], or the true spirit dephlegmated.

This spirit or oil put upon iron[s24] transmutes it in a little time into copper[s10].

(Menstruum 13) Spirit of Mined Nitre[s1] Of Tentzelius

Rx. Hungarian mined nitre[s1] calcined white (in the sun with a burning glass, called philosophic calcination) 1 pound, potassium bitartrate[s21], calcined black, a half-pound; add to them (being in powder) SV. Distil in an alembic with a strong fire, cohobate it, and separate the SV. from the spirit of mined nitre[s1] [43], by rectifying it in sand.

---

[43] The 1904 text, page 25, shows the symbol as "s17" shown in the Symbol Dictionary, which I believe to be an error made by the typesetter.

OF URINE[s20]

Of this is prepared some very rare dissolvents of singular force. I include here only those needed in our praxis.

(Menstruum 14) Spirit of Urine[s20]

Rx. Fresh or new made boy's urine[s20] that drinks wine, distil by alembic in BM., cohobate it, and you have phlegm and spirit; separate this according to art (which is done by a small retort in sand), and elevate the spirit in a glass body, so shall it be very volatile and white, but exceeding stinking. This is a notable lithontriptic[44] (the meaning is: "") and will dissolve the stone if injected into the bladder with a proper syringe.

(Menstruum 15) Oil of Urine[s20]

Rx. Of that gritty or tartarous matter which adheres to the bottom and sides of the urinal, calcine, dissolve, coagulate and then dissolve again per deliquium. This, if given 1 scruple[s12] in a convenient vehicle, perfectly dissolves the stone.

---

[44] Having the quality of, or used for, dissolving or destroying stone in the bladder or kidneys.

(Menstruum 16) Volatile Salt[s5] of Urine[s20]

Rx. The urine[s20] of a boy or young man, SV., in equal parts, mix and evaporate to the consistency of a new honey; put it into a long-necked glass, and distil it with so small a heat in ashes or sand that it may condense in the alembic, and there will come forth in the alembic a white spirit like snow, which in the cold will coagulate. If this spirit be joined with the salt of the fæces, and volatilized by often cohobations, it will be a notable menstruum to draw the vitriol of metals, chiefly of silver[s9]; if it yet be digested with common salt[s5], and purified by often solutions and coagulations for about ten days and nights in BV., it will resolve; and by the addition of rectified SV. and ten days' digestion, it will be a good menstruum to dissolve gold[s15].

OF VINEGAR[s22]

That of wine is best, and yet better if vitriolated. The modern chemist is of the opinion that alcohol cannot be acetified, if any essential oil of pyroligneous acid[45] is present. Vinegar[s22] of metheglin is best for the dissolution of metals, for it has both an animal and a vegetable spirit, and so has the greater power of dissolution, and is therefore called Philosophic Vinegar[s22].

(Menstruum 17) Distilled Vinegar[s22]

Rx. Put it into a glass still in BM. or ashes with a gentle heat, draw off the phlegm without taste, which will be near a quarter part; change the receiver and force over the spirit. If it be for physical uses, you must take heed of burning it lest your extracts smell of it. If for metals, draw off the phlegm in a gentle heat in BM. then in sand distil violently until a red spirit ascends, and all be come over.

(Menstruum 18) Spirit of Vinegar[s22] of Clossaeus.

Rx. Six quarts, and distil in a luke-warm BM. till but a quart remains; then in ashes draw off the remainder to dryness, cohobating two or three times upon the fæces. If you would have it yet stronger, make balls of the crystals and bole, and drive it

---

[45] Also called 'wood vinegar'; formed from dry ditillation of wood; mostly acetic acid + methanol.

through a retort, so have you spirit which rectified will be fiery and corrosive.

(Menstruum 19) Radicated or Alcalized Vinegar[s22] of Sennertus[46].

Rx. Fæces of distilled Vinegar[s22], calcine them dry, then add distilled vinegar[s22], and draw it off again in sand, cohobating so often till all its common or fixed salt[s5] ascend with the vinegar[s22].

---

[46] Daniel Sennertus was an alchemist of the 17th century.

OF SV.[47]

As has been intimated, the SV. of commerce is generally lacking in potency by reason of its aquosity, and it is best here, as in all the foregoing preparations, to have resort to one's own laboratory. One of the best ways to test its strength and purity is to dip a cotton wick like that of a candle, and setting it on fire; if the flame fasten on the wick, it is a sign of the goodness of the spirit; but if it does not, it is weak and not sufficiently dephlegmated. That known as Canary[48] is best for these purposes.

(Menstruum 20) Common SV.

Rx. Distil by a vesica till the finest parts are ascended (which is known by the taste); let it be several times rectified in BM. drawing off the half, third or fourth part, till the spirit is high, and no humidity will remain that will flame. Where note, that the orifice of the still being covered with a four-fold thin paper or thick cloth, the spiritual parts only will penetrate, and the watery fall back again; and if you have a still with a long neck or serpentine glass (worm), you will rectify it the better[49].

---

[47] Spiritus Vini; spirit of wine.

[48] Canary, or sack, is an antiquated wine term referring to white fortified wine imported from mainland Spain or the Canary Islands.

[49] Hans Nintzel noted that a Kjeldahl bulb is excellent for this.

(Menstruum 21) Tartarized SV.

Rx. Of SV. 2 pounds, potassium bitartrate[s21] in powder 1 ounce[s11]. Distil in a bath full of sawdust or straw moistened with water, so that the drops may fall leisurely, which cease when the phlegm begins to come; but if the phlegm ascend with it, rectify it. Before rectification, you may repeat the former work, with the addition of an ounce more of potassium bitartrate[s21]. Sennertus says, if you rectify the spirit upon the same potassium bitartrate[s21], with several cohobations, making a strong fire at the conclusion (casting away the phlegm that comes in the middle), you shall have SV. tartarized.

Boyle[50], the experimental philosopher and Hermetic chemist of the XVllth century, gives the following process, which is to be recommended, as it dispenses with frequent rectification: "Put about an inch thick of potassium bitartrate[s21] calcined to whiteness (for I find it not necessary to reduce it to a salt) and very dry into the bottom of a tall and slender glass body, and pour on it as much SV., but once rectified, as will, when they have been shaken together, swim above the potassium bitartrate[s21] a fingers breadth, and then the head and receiver being carefully fastened on again, in a gentle heat draw off the SV., shifting if you please the receiver when about half is come over, and if need be rectifying once more all that you distil upon dry calx of potassium bitartrate[s21] as before.

---

[50] Robert Boyle FRS (1627–1691) was a 17th century natural philosopher, chemist, physicist, and inventor, who is also noted for his writings in theology. Boyle is generally considered the first modern chemist; Boyle's Law is foundation of physical chemistry. His *The Scyptical Chemist* is a cornerstone book in the development of chemistry.

Therefore this alcohol of SV. we peculiarly call the Alcalizate SV., and the rather because SV. Tartarizatus, which perhaps may be thought the most proper name for it, is employed by eminent chymical writers to signify a different thing.

(Menstruum 22) Spiritus Ammonium Chloride[s13]

Rx. Ammonium chloride[s13] 1 part, ashes 4 parts, distil[s16] by an alembic in sand, so have you a spirit, which rectify in a long still. Or thus: quench red hot brickbats in the solution of ammonium chloride[s13] in fair water[s4], till such time as all the water[s4] is drunk up, then distil[s16] in a retort. Or thus: which makes a wonderfully piercing spirit due to the vol. salt[s5] of urine[s20]: Impregnate water[s4] with ammonium chloride[s13], as much as it will hold, in which imbibe leaves of brown paper made into balls; put them in a still and with sand or a reverberatory draw off an acid spirit of a golden color, which rectify until it is white or clear. Or thus: Mix the salt[s5] with clay and make it into balls, then distil[s16] in a retort as with common salt[s5].

John Hazelrigg

Seven Universal Medicines of the Ancients

While there were many medicinal preparations of the Spagyric chemists and philosophers which they termed powerful and universal, there were seven particular ones, enumerated as follows: Aurum Potable, Tinctura Auri, Precipitatus Aureus, Aurum Vitae, Hercules Bovii, Manna Mercurii and Bezoarticum Solare. These were accounted notable remedies in the treatment of most if not all diseases, and are here given in full and demonstrable manner. Different methods of procedure were generally observed by different authors, some of which were very complex and obscure, but the analogies and purport of the various processes were obvious, and the results found to be identical in all cases.

(1) AURUM POTABLE

Rx. Dissolve leaf gold[s15] 1/2 ounce[s11] in aqua regia (Menstruum 4), precipitate it by the affusion of oleum potassium bitartrate[s21] per deliquium (Menstruum 11), so will it look white like lime; then wash off the salts with common water[s4], and dry the quicklime[s25] gently by itself or in the sun, for if dried in an oven heat it is liable to explosion. In this form it is Aurum Fulminans[51].)

Take of this quicklime[s25] (reverberated to the highest brownness and porosity, or often calcinated[s26] with aqua regia), digest[s2] in spirit of urine[s20] (Menstruum 14) with a gentle heat, in a close luted

---

[51] (Gold fulminate, $ClAuNH_2)_2NH$ or (even worse) $OHAuNH_2)_2NH$ is <u>highly</u> explosive and extremely dangerous.

35

vessel, for a month, or until there is a red tincture like blood; decant and add fresh spirit and repeat this work till there is no more tinged. Put the solutions together and digest[s2] for 20 days or a month; then with a gentle heat in BM. separate the spirit or menstruum (to be kept for the same use), and at the bottom there will be left a red tincture like oil, which will dissolve in any liquor, and is Aurum Potable.

Dose, in sack[52] or SV. 1 scruple[s12] at a time for several days, or may give 4 to 8 drops in any other vehicle. Paracelsus says 1 scruple[s12]. This is a strong sudorific.

Rx. Another way: quicklime[s25] of fine gold[s15] made by aqua regia (Menstruum 2 or Menstruum 4) and perfect reverberation, add volatile salt[s5] of urine[s20] (Menstruum 16), which digest[s2] 40 days in a gentle and equable heat in a vessel close luted, and the menstruum will become blood-red; decant and reiterate with fresh menstruum, and again repeat, put the solutions together and digest[s2] for a philosophical month, and proceed as in the above.

Aurum Potable Clossaei, Potable Gold of Clossaeus.

Rx. Gold[s15] 1 ounce[s11], mercury[s23] 6 ounces[s11] make an amalgam, which beat with as much common salt[s5] melted; evaporate the mercury[s23] with a gentle fire[s8]; wash away the salt[s5] with warm water[s4]; beat the remaining quicklime[s25] with common sulfur[s27] [53] 3 ounces[s11] which cement 3 or 4 hours in a crucible with a hole at the top, till all the sulfur[s27] is vanished. Repeat this seven times with

---

[52] Sack is an antiquated term referring to white fortified wine imported from mainland Spain or the Canary Islands.

[53] The symbol used here is the inverted symbol for sulfur and appears to be a typesetter's error.

fresh sulfur[s27], till the gold[s15] be like a red sponge, which sublimate[s28] with eight times as much ammonium chloride[s13], till it be of the color of sandarach[54], which sweeten by washing; then take SV. 13 parts, spirit of salt[s5] of potassium bitartrate[s21] (Menstruum 10b) 1 part, mix them, and distil[s16] together; into which put either the aforesaid spongy quicklime[s25], or quicklime[s25] sublimed, and digest[s2] till it is dissolved. If you abstract the SV. you will have a yellow powder. The virtues of this are almost innumerable.

Aurum Potable Quercetani, Potable Gold of Quercetan.

Rx.   A light spongy quicklime[s25] of gold[s15], from which with spirit of vinegar[s22] (Menstruum 18) draw a tincture by digestion[s2] in BM. which exalt by circulation with SV. This is also called Aurum Vitae and is, as Quercetan says, of incredible virtues for almost innumerable diseases.

Aurum Potable Grulingii, Potable Gold of Grulingius.

Rx.   Quicklime[s25] of gold[s15] made by reverberation to the highest tenuity, upon which put menstruum made of equal parts of volatile salt[s5] of urine[s20] (Menstruum 16) and rectified SV. (Menstruum 20) digested 12 days in BM. with an equal and gentle heat; being mixed in a glass vessel, seal up the quicklime[s25] and menstruum hermetically, and digest[s2] them 40 days, till the tincture rise red as blood, which work often repeat. Put these general tinctures together and digest[s2] eighteen days, then by distillation[s16] with a gentle heat separate the spirit, and the gold[s15] will remain in the bottom of the

[54] A resin obtained from the small cypress-like tree *Tetraclinis articulata* of northwest Africa.

glass in the form of a moist red oil. This solution distil[s16] by a glass retort in sand so often till the tincture of gold[s15] come over of a most blood-red color, and there remain in the bottom nothing but a dry, spongy, black earth.

There are many other preparations of Aurum Potable, but these will suffice here. This formula of Grulingius is of especial virtue, is used as a universal remedy, for it restores and preserves the radical humidity both in quantity and quality entire, and frees the powers of the whole body from the malignity of diseases, keeping it safe from corruption during the term of the natural life. It is excellent in apoplexies, epidemical diseases, pestilent fevers, palpitation of the heart; it provokes the terms, causes speedy delivery in child-birth, yet prevents miscarriage; takes away the malignity of cancers, causing their speedy healing; restores in consumptions. These medicines are strongly sudorific, causing a sweat that carries away with it the seeds and roots of malignant and poisonous afflictions. It is in vain to enumerate in particular all they are capable of doing, so universal are their curative faculties.

(2) TINCTURA AURI

Tinctura Auri Basiliana, Basil Valentine's Tincture of Gold[s15]

Rx. Of the quicklime[s25] of the most fine gold[s15], made by dissolution in Basil's aqua regia (Menstruum 5), which volatize with spirit of salt[s5] (Menstruum 8) actuated with spirit of ammonium chloride[s13] (Menstruum 22); precipitate with oil of potassium bitartrate[s21] (Menstruum 10a), or by a gentle abstraction reverberate with flowers of potassium bitartrate[s21], then with rectified SV. and spirit of salt[s5] extract the sulfur[s29] of gold[s15], which digest[s2], and abstract gently the spirits; then dissolve[s3] it again in the aforesaid aqua regia and volatize with SV.

This is a Golden Liquor of great force, having all the virtues of Aurum Potable. Another formula, given by Quercetan, and which he says is of almost incredible virtues for innumerable diseases, is thus:

Rx. Spongy light quicklime[s25] of gold[s15], draw a tincture by digestion[s2] in BM. with spirit of vinegar[s22] (Menstruum 18), which exalt by circulation with SV. Another formula given by Salmon:

Rx. Aurum Fulminans (see first formula for Aurum Potable), well sweetened, 1 part, volatile spirit of mined nitre[s1], 4 parts.; digest[s2] in a warm bath for 40 days or more; decant the red-tinged spirit, and evaporate to dryness; put to it SV. (Canary) with white potassium bitartrate[s21], and in a gentle heat extract to the color of a ruby, which distil[s16] and bring to a consistency like powder; wash of the SV. with distilled[s16] water[s4], and you may dissolve[s3] it in any convenient vehiculum. Note, if you have not the aforesaid spirit of mined nitre[s1], you may use this:

Rx. Salt[s5] 1 pound nitre 6 ounces[s11] distil[s16] a spirit, which mix with equal parts of SV. and draw off the mixtion[55] from a great alembic.

(3) PRECIPITATUS AUREUS

Rx. Leaf gold[s15], or filings of fine gold[s15] 1/2 ounce[s11], dissolve[s3] it in aqua regia without ammonium chloride[s13]; take glass of antimony[s19] 1/2 ounce[s11], dissolve it in aqua fortis (Menstruum 1); cleansed mercury[s23] 3 ounces[s11], dissolve it also in aqua fortis; mix all the solutions and draw a water[s4] by an alembic; then add fresh aqua regia, draw off the same often, till the precipitate fumes not when laid on a red hot iron; calcine[s26] all, that the aqua regia may be spent; then distil[s16] from it SV. six times and calcine[s26] the matter gently.

It purifies the whole mass of blood and the whole body; it cures jaundice, scurvy, dropsy, gout; it provokes urine, dries up all moist humors, and opens all obstructions; it cures the epilepsy, colic, quartan, and all cancerous and malign ulcers. Dose, 4 to 5 grains. This called the Golden Precipitate.

(4) AURUM VITAE

Aurum Vitae Sennerti, Aurum Vitae of Sennertus

Rx. Mercury[s23] purified 5 ounces[s11], fine plates of gold[s15] 1/2 ounce[s11], make an amalgam, and wash it with vinegar[s22] till all the blackness be gone; then put it in a retort and add aqua fortis (Menstruum 1) one pint, digest[s2] in ashes or sand, that the

---

55 mixture (archaic)

mercury[s23] and gold[s15] may be at the bottom in a powder; then distil[s16] and increase the fire[s8] at the end, that the bottom of the retort may be red hot, and all the corrosive spirits come forth; the vessel being cooled, powder the matter and pour on the abstracted aqua fortis again, and cohobate, and there will be a red powder at the bottom; which keep, casting away what was sublimed at the sides of the retort; then heat an iron red hot and sprinkle on it the mercurial[s23] powder, not only to dry it but to evaporate what is volatile, then keep it in a closed glass vessel.

Sennertus says: This is one of the most noble medicines yet this day known in the world, and will do as much as any whatsoever. It will easily, safely, and speedily cure any old, malign, and deplorable disease. It is a wonderful Arcanum in the dropsy, pox and gout, as also in the jaundice, all manner of defluxions, scurvy, leprosy, scabs, itch, plague, poison, all fevers, and all obstructions in any part of the body. It begins, continues and perfects the cure alone. It is a great diaphoretic, and may be given in a dose, 3 to 6 grains; if to purge in purging pills, but if to sweat, in some cordial essence, elixir or electuary.

Aurum Vitae Hartmanni, Hartmann's[56] Living Gold, or Gold of Life.

Rx.   Filings of gold[s15] 1 ounce[s11] dissolve[s3] it in aqua regia 4 ounces[s11], keep the solution hot; take mercury[s23] 12 ounces[s11], dissolve[s3] it in a pint and a half of aqua fortis communis, mix them till they are black, distil[s16] in an alembic in sand with a gradual fire till the still and that at the bottom is red hot; then calcine[s26] it with a red hot iron, till the spirits of the aqua fortis are gone, wash it with

---

[56] Franz Hartmann was an alchemist of the 19th century.

41

water[s4], and distil[s16] SV. often from it by cohobation, so have you Aurum Vitae.

NOTE: the aqua regia for this work is thus made:

Rx. Aqua fortis of mined nitre[s1], nitre[s6], salt[s17], 4 ounces[s11] of each, ammonium chloride[s13] 1 ounce[s11], distil[s16] them in sand in an alembic. It has the virtues of the above; dose, the same.

(5) HERCULES BOVII

Hercules Bovii, Hercules of Bovius[57]

Rx. Filings of fine gold[s15] 1 ounces[s11] mercury[s23] cleansed 4 ounces[s11] dissolve[s3] both asunder in aqua regia, distil[s16] both together with a gradual fire, and after in the end with a greater, in a retort; then with fresh aqua regia dissolve[s3] the precipitate at the bottom and the sublimate in the neck of the retort, and distil[s16] so long till all become a precipitate; then calcine[s26] it on a red hot iron to fix all the corrosive spirits, and sweeten it by ablution in SV.

Bovius saith, It is the best of all purges; it kills all worms, cures the French disease, smallpox, plague, leprosy, quartans, and many other diseases otherwise incurable. Dose, 3 to 4 grains with sugar of violets, broth, or in purging pills.

(6) MANNA MERCURII

Manna Mercurii, or Golden Panchymagogon

Rx. Of mercury[s23] dulcis, as much as you like, elevate it by often sublimation[s28], till it turns to fixed crystals, which dissolve[s3] into a liquor; of which take 6 drams[s14] and of gold[s15] calcined[s25] 2 drams[s14] mix and digest[s2] for 40 days; cohobate till it melt in the still like wax. Mercury Dulcis is the sweet sublimed Mercury, or Tamed Dragon, of Quercetan.

---

[57] Astrologer, physician and alchemist of the 16th century.

It is a great specific and secret against the French-pox, and all manner of venereal evils. By this mercury[s23] is brought to the highest degree for physic, and is made of wonderful virtue for curing vertigos, megrims, and other diseases of the brain. It is in no ways inferior but contains all the virtues of Aurum Vita. It is a good sudorific, and cures chiefly by sweating. Dose, 3 to 8 grains.

(7) BEZOARTICUM SOLARE

Bezoarticum Solare, or Solar Bezoare

Rx. Tincture of gold[s15] (extracted from the oil-like solution of gold[s15], and sweetened by abstracting often from it strong vinegar[s22]) 2 ounces[s11] butter of antimony[s19] dissolved[s3] in spirit of salt[s5] (Menstruum 8) 14 ounces[s11], mix them, unite by cohobation, then calcine[s26], where note that the tincture of gold[s15] is extracted with the menstruum of Basil (Menstruum 5), digesting[s2] them a month. Thus Tentzelius.

Or thus: Rx. Butter of antimony[s19] 1/2 pound dissolve[s3] it by pouring on it gradatim spirit of salt[s5] (Menstruum 8); then take fine leaf gold[s15] 1/2 ounce[s11] dissolve[s3] it in aqua regia (Menstruum 2), mix both, and abstract the menstruum by degrees, and pour it on again; add fresh spirit of salt[s5], abstract, reiterate it often, wash, dry and fire the quicklime[s25] with rectified SV. Thus Crollius[58].

---

[58] Oswald Croll (1560 – 1608) was a professor of medicine and alchemy at the University of Marburg in Hesse, Germany. A strong proponent of alchemy and using chemistry in medicine, he was heavily involved in writing books and influencing thinkers of his day towards viewing chemistry and alchemy as two separate fields.

Or thus: Rx. Spiritual gold[s15] (spiritualized by the bezoardic spirit of salt[s5] (Menstruum 7)) 1/2 ounce[s11], dissolve[s3] it in aqua regia (Menstruum 2); dissolve[s3] in the same butter of antimony[s19] rectified 4 ounces[s11] or 6, mix both, abstract the menstruums by a retort by often distillations; then with a gentle calcination[s26] bring it to a violet colored powder, or with a strong calcination[s26] to a purple, which is better than the former. Thus Schröder.

Or thus: Rx. Spiritual gold[s15] (ut supra), add to it butter of antimony[s19] dissolved[s3] in spirit of salt[s5] or in aqua regia; unite and fix them by often distillations; then abstract, and by calcination[s26] you will have a Bezoardic gold[s15] of purple color and great force. Thus Hartmann .

This is a wonderful medicine, and is the seventh Medicament in name, number and nature of those which may be called powerful and universal; it performs all that the others will do. It is a great sudorific, and may be given from 2 grains to 10 grains.

John Hazelrigg

CHOICE SPAGYRIC PREPARATIONS

Abstracted from the Teachings of the Ancients, and
Transcribed into Clear Formulas of Practice.

THE VULNERARY OF MINED NITRE[s1]

Rx. Pure rectified oil of mined nitre[s1] (Menstruum 12) 1 ounce[s11].
SV. rectified, 2 pounds. Mix and digest[s2].

This is the greatest secret in mined nitre[s1]. It cures most diseases
of the head, as the megrim, epilepsy, apoplexy, vertigo, etc., and is a
wonderful thing in all manner of sores, ulcers, cancers, and the like,
and cures green wounds at one dressing. Inwardly it cures coughs,
colds, asthmas, ulcers of the lungs, consumptions, pleurisies, stone
and gravel in the reins and bladder, and all sorts of fevers, whether
continual or intermittent. It opens all obstructions of the stomach,
bowels and kidneys, purifies and sweetens the mass of blood, cures
the scurvy, French-pox and other ill-habits of the body. Dose, 2
drams[s14] to 4 drams[s14] in any proper vehicle.

THE STAR OF MERCURY[s23] (Stella Mercurii.)

Rx. Mercury[s23] seven times sublimated[s28] and as often rectified
with quicklime, with a gentle heat dissolve[s3] it in spirit of salt[s5]
(Menstruum 8), abstract the spirit, sweeten and boil the mercury[s23]
in distilled vinegar[s22] (Menstruum 17), and wash it with distilled[s16]
rain water[s4], dry it, and digest[s2] it in SV., which with a gentle heat

47

drive through a retort, increasing the fire[s8] (what remains keep to make a salt[s5] of), abstract the SV. in BM. and you shall have at the bottom a fragrant sweet oil, which according to Basil is the Star of Mercury[s23]. The salt[s5] of mercury[s23] is made thus:

Rx. Of the body that remains after the preparation of the Stella Mercury[s23] put upon same its fragrant oil, digest[s2] and extract the salt[s5] of mercury[s23]; to the decanted extraction put SV., digest[s2], and abstract in ashes; and after it you shall have an oil of mercury[s23], and the salt[s5] at the bottom, which possesses all the virtues of the oil.

It is a diaphoretic, and is a proper specific in venereal troubles, in which it is a great secret, though it be ever so old; it cleanses the blood and cures all scabs, tetters, and ulcers, although old and malign. Dose, of the oil, 4 grains; of the salt[s5], 1 to 3 grains.

THE ALBION POWDER (Pulvis Anglicanus)

Rx. Of the best antimony[s19] as much as you like, calcine[s26] it alone; then take of the aforesaid quicklime[s25] and nitre[s6] 1 pound each white potassium bitartrate[s21] 1/2 pound; mix and calcine[s26] till the detonation is over; repeat this last work again, and the third time, adding nitre[s6], potassium bitartrate[s21], and sulfur[s29], in equal parts; lastly, wash it with hot water till it is sweet.

It is accounted a universal medicine against all diseases; it opens all manner of obstructions, provokes the terms, cures surfeits, colic, small-pox, all sorts of agues and fever whatsoever, gout, dropsy, etc. Dose, 1/2 dram[s14] to 1 dram[s14].

TO VOLATIZE GOLD[s15]

Rx.   Dissolve[s3] leaf gold[s15], or filings of gold[s15], in the Bezoardic Spirit of common salt[s5] (Menstruum 7), rectified oil[s7] of salt[s5] (Menstruum 8), aqua regia, or oil[s7] of antimony[s19], by a gentle heat (lest the spirits should ascend too fast), abstract the menstruum and add fresh, repeat four or five times, till it is oily; then distil[s16] this solution in a glass retort with a strong fire[s8], to force away the spirits, cohobate and repeat this work till the gold[s15] be blood-red. Lastly, having freed the solution from all corrosive spirits, cohobate with SV., wormwood, or any other vegetable spirit, and distil[s16] by an alembic; let the cohobation and distillation[s16] be often repeated and the gold[s15] will come over in a liquid form, and have all the virtues of the most exquisite Aurum Potable.

THE ANODYNE MAGISTERY, OR
SULFUR[s29] OF MINED NITRE[s1]

Rx.   Mined nitre[s1] of iron[s24], or the best Hungarian mined nitre[s1] 1 pound. Dissolve[s3] it in a sufficient quantity of rain-water, which filter; then take filings of pure iron[s24] or steel, and often moisten or sprinkle them with aforesaid mined nitre[s1]. Fire[s8], as often drying them with a very gentle heat, repeating this so long till the filings are reduced to a rubicund mass, which beat into a subtle powder, upon which affuse spirit of vinegar[s22] (Menstruum 17), so much as may overtop it the breadth of five fingers in a great and large glass. Digest[s2] with a sufficiently intense heat till the vinegar[s22] is tinged, which, whilst warm, decant; affuse more fresh spirit, which repeat so often till the new affused vinegar[s22] will be no more tinged red. Mix these solutions or tinctures together, and with potassium bitartrate[s21] per deliquium (Menstruum 11) precipitate the yellow sulfur[s29] of mined nitre[s1], from which decant the supernant liquor, and edulcorate with many affusions of fair water[s4] warmed, till there be found no taste either of vinegar[s22] or mined nitre[s1] to remain; which then dry, put it into a bolthead[59], seal it up hermetically, and in sand calcine[s26] it to redness, which will be done in a short time.

NOTE: If you would have the sulfur[s29] inflammable, you must instead of common mined nitre[s1] take the mined nitre[s1] of iron[s24], prepared according to art, and use it as aforesaid upon the filings of steel, not precipitating it with oil of potassium bitartrate[s21], but evaporating in BM.

From this Magistery or sulfur[s29] with SV. and salt[s5] of potassium bitartrate[s21] is prepared a noble medicament and essence of so great virtues that they exceed the sphere of my commendations; performing all and more than any preparation of opium can do, and with far greater safety. It is a perfect cure for the falling sickness, vertigo, madness, melancholy, gout, and other chronic and radicated diseases. Dose, 6 to 10 grains or 12.

MAGISTERY of MINED NITRE[s1]
(According to Sennertus)

Rx.   Spirit of mined nitre[s1] rectified with the oil (Menstruum 12); with the phlegm draw a salt[s5] from the caput mortuum[s18], which separate from the salt[s5] by Menstruum 16 , often cohobating and digesting. Take of this 2 parts, of the aforesaid spirit and oil 1 part mix, and exhale the humidity in BM., which repeat so long till the

---

[59] An illustration of a bolthead is shown on the following page.

Bolthead, from *The Art of Distillation* by John French

    A. The crooked pipe

    B. The glass body

    C. The glass stopple

    D. The mouth of the vessel itself

salt[s5] has sucked in its equal weight of oil. Then decant in a luted glass matrass for eight or ten days; lastly coagulate in sand or ashes, which will be done in sixteen or twenty days.

This is a great diuretic, breaks and expels the stone, opens all manner of obstructions, chiefly of the spleen, cools the heat of fevers, and cures dropsy, scurvy, etc. Dose, 4 to 6 grains.

OIL OF THE SULFUR[s29] OF MINED NITRE[s1]

Rx. Sulfur[s29] of mined nitre[s1] 4 ounces[s11]. Salt[s5] of potassium bitartrate[s21] 2 ounces[s11], mix and distil[s16] by a retort, at last make a fire[s8] of suppression, so have you a red oil. If you drop spirit of vinegar[s22] (Menstruum 18) upon this oil, you have a pectoral powder, which edulcorate and dry.

Both oil and powder open obstructions of the lungs, and cure almost all diseases afflicting those parts; externally it cures wounds and ulcers. Dose, 5 to 10 grains. From this powder you may make an essence or tincture, for which see following.

ESSENCE OF THE SULFUR[s29] OF MINED NITRE[s1]

Rx. Precipitate of the oil of the sulfur[s29] of mined nitre[s1] (ut supra), digest[s2] in SV. in a hot place in a close vessel for eight or ten days, so the essence will swim at top like oil, which filter[s30] from the SV.

This Essence is sweet, and according to Hartmann is of as great power and force as the Tincture of Antimony. Given with essence of balm and choice canary wine it doth wonders in the art of healing. It expels all bad humors by sweat, cures dropsies, consumptions, and the stone; it strengthens the womb, takes away barrenness, and causes fruitfulness in both sexes. Dose, one to four grains.

THE ENS OR BEING OF COPPER[s10]
(Ens Veneris)

Rx. The red caput mortuum[s18] of aqua fortis extract all its salt[s5] with fair water[s4] , which dry and powder; pour thereupon the spirit of ammonium chloride[s13] (Menstruum 22) or urine[s20] (Menstruum 14) ; stir it and dry it well, repeating this work seven times; then powder it and mix it with purified ammonium chloride[s13] 2 parts; grind and mix them well, and in a glass retort in sand, by degrees of fire[s8], sublimate[s28] for half a day, stopping the mouth of the retort with cotton or wool, and in the neck of the retort you will have the Ens Veneris of a yellow or gold-like color, which you may mix with its equal quantity of colcothar or caput mortuum[s18], and sublimate[s28] once or twice more.

It is a noble and worthy anodyne, easing all manner of pain, and causing rest. It cures the rickets, and kills worms in children, performing those things beyond any other medicine. It has no equal in pleurisies and in suffocation of the womb; it opens all obstructions, and exhilarates the heart, comforts the animal spirits, gives ease in the stone and dissolves it, and cures such as are in consumption. Dose, 4 to 8 grains, or 10 or 12 in canary wine or other convenient vehicle.

THE FAMOUS ELIXIR OF LIFE

(Prepared from Balm)

In the proper season of the year, when the herb is at its full growth, and, consequently, its juices in their whole vigor, gather at the fittest time of the day (when Jupiter is rising, and the Moon in Cancer is applying to a conjunction, sextile or trine aspect thereto) a sufficient quantity of balm, wipe it clean, and pick it; then put it in a stone mortar, and by laborious beating reduce it into a thin pap.

Take this glutinous and odoriferous substance and put it into a bolthead, which is to be hermetically sealed, place it in a dunghill, or some gentle heat equivalent thereto, where it must digest[s2] for forty days. When it is taken out the matter will appear clearer than ever, and have a quicker scent. Then filter[s30] the grosser parts, which, however, are not to be thrown away. Put this liquid into a gentle bath that the remaining gross particles may perfectly subside. In the meantime dry, calcine[s26], and extract the fixed salt[s5] of the grosser parts (which remained after the above filtration[s30]), which fixed salt[s5] is to be joined to the liquor when filtrated.

Next, take sea salt[s5], well purified, melt it, and, by setting it in a cold place, it will dissolve[s3] and become clear and limpid. Take of both liquors equal parts, mix them thoroughly, and having hermetically sealed them in a proper glass, let them be carefully exposed to the sun, in the warmest season of the year, for about six weeks. At the end of this space the primum ens of the balm will appear swimming at the top like a bright green oil, which is to be carefully filtered[s30] and preserved.

Of this oil, a few drops taken in a glass of wine for several days together, will bring to pass those wonders that are reported of the

Countess of Desmond[60] and others; for it will entirely change the juices of the human body, reviving the decaying frame of life, and restoring the spirits of long-lost youth. The author who records this curious discovery remarks: "If after the medicine is thus prepared any doubt be had of its efficacy, or of its manner of operation, let a few drops be given every day on raw meat to any old dog or cat, and in less than a fortnight, by the changing of their coats and other incontestable changes, the virtue of this preparation will sufficiently appear."

This is of the nature of a Quintessence - being similarly prepared, - the alchemical praxis for which will be fully and clearly expounded in a work I now have in preparation.

THE PHILOSOPHER'S WATER
(Aqua Philosophorum)

Rx. Calcinate[s26] potassium bitartrate[s21] till it be greenish-blue or sky-colored, pour on SV. tartarized (Menstruum 21), digest[s2], then distil[s26], and at last force it with a violent fire[s8].

This is wonderful in curing diseases arising from tartar, in the scurvy, quartans, melancholy, asthma, dropsies, and obstructions of the liver, spleen and bowels. It is the best menstruum to make all purging tinctures and extracts, whether out of vegetable or mineral. If it be circulated in a bolthead hermetically sealed, it becomes balsamic and sweet-scented, and from a crystalline color it becomes that of a ruby, being as it were a Balsam of Life and Vital Powers, exalting Nature to her highest degree of purity and clarity by quickening the internal fire and heat. Three or four drops of it given

---

[60] Possibly Katherine FitzGerald, Countess of Desmond (d. 1604).

with essence of saffron gives ease and rest, and restores in consumptions.

ARGENTUM POTABLE
(Potable Silver of Clossaeus)

Rx. Plates of silver[s9] which calcine[s26] often with sulfur[s29], pour on water[s4] and set it to shoot into crystals; dissolve[s3] them into a spirit, by rectified SV. (Menstruum 20) actuated with spirit of salt[s5] (Menstruum 8) and spirit of nitre[s6] (Menstruum 6); digest[s2], then filter[s30], and bring the azure or blue tincture to a powder, which dissolve[s3] in rectified SV.

This is a wonderful and excellent medicine against the epilepsy, and most other diseases of the head, as the Moon bears astrological rule over the brain. You may also make Potable Luna after the method of Clossus in that of gold[s15], already given.

SPIRIT OF SILVER[s9]
(Spiritus Argenti)

Rx. Of the crystals of silver[s9] (ut supra) being twenty times calcined[s26] (that it may yield its spirit the easier), or you may take crystals made of filings of fine silver[s9], with 3 ounces[s11] of water of lilies of the valley, actuated spirit of mined nitre[s1] of copper[s10] 1 ounce[s11] , digest[s2] 40 days in ashes, till the menstruum is of a greenish blue; filter[s30] this, and add more spirit of mined nitre[s1] of copper[s10] , and extract till the silver[s9] is nearly all dissolved[s3]. Put the solutions or tinctures together, evaporate and crystallize, which dry; put these dried crystals into a retort, from whence draw first a phlegm and spirit of copper[s10], which keep apart; then with a stronger fire[s8] force over the spirit of silver[s9], and lastly an oil.

It is a thing found by experience that this Luna Spirit takes away the falling sickness by the roots; it specifically strengthens the head and comforts the animal spirits. It is good against palpitation of the heart, madness, and all melancholic distempers. The oil taken 3 or 4 drops in balm, sage, rosemary, or peony water, is more effectual to all the purposes aforesaid.

OIL OF CINNAMON

Rx.   Take 1 pound of grossly bruised cinnamon, which cover with SV. made very sharp with spirit of salt[s5] (Menstruum 8), or else having as much salt[s5] put into it as the SV. will dissolve[s3]. Put them into a blind head, which lute close, and set it to digest[s2] in a gentle heat for about ten days, then apply an alembic close luted with its receiver; distil[s16] it with a small fire[s8] by degrees, so shall you have a heavy oil, which will sink, and a spirit, which filter[s30] by setting in a cold cellar for 14 or 16 days after they are distilled[s16], by which time the oil will settle to the bottom.

NOTE: If this spirit, after its filtration[s30], be joined to its own proper salt[s5], or else salt[s5] of potassium bitartrate[s21], and after they are sufficiently united by digestion[s2], circulated with its own proper chemical oil so long till all become united and one entire body, so have you an elixir. This in a true sense is a fortified quintessence, for it is a unition of the three principles, salt[s5], sulfur[s29], and mercury[s23], together with the essence. 1 pound yields but 2 drams[s14] or little over, of oil.

This oil pierceth even the flesh and bones, being very hot and dry, and is good against all cold and moist diseases afflicting the head, heart, and other principal parts, in so much that if one lay speechless and almost breathless, it would presently recover him. It helps all diseases that come from cold and phlegm; it digests, makes thin, and provokes the terms, and brings away both birth and after-birth; it helps coughs and asthmas, and stops all fluxions from the head and brain. It is one of the greatest vegetable cordials, and perfectly cureth consumptions, comforting nature, reviving the heart, and cheering all the spirits, natural, vital and animal. Dose, 2 to 10 drops in any convenient liquor. Some give it in cordial waters, some in broth, some in milk, and some in canary wine. The best way to take chemical oils is to drop the intended quantity on a piece of refined or loaf-sugar, letting it soak into it, and then dissolve the sugar in wine or some cordial water, proper to the distemper.

ELIXIR OF PROPERTY

Rx. Of aloes, myrrh, saffron, each 1 ounce[s11] moisten all with tartarized SV. (Menstruum 21) and bring them to an alcohol (see Note on Boyle, Menstruum 21); put all into a glass body, with a pint of tartarized SV., and so much oil of sulfur[s29] *per campanum*[61] as may flow two or three inches above all; close it well and circulate all for three months, extract the tincture and decant it; add tartarized SV. to the matter remaining, extract again and decant it; then distil[s16] the fæces that remain, and add it to the former; and again for a month (without distillation[s16]) circulate them. Others extract the tincture with salt[s5] of potassium bitartrate[s21] volatilized with SV. by digestion[s2], so long till the bitterness of the aloes is not perceptible, which seems to be a better way than the former.

---

[61] Made by means of a glass bell-like device (Hans Nintzel).

This is the famous Elixir Proprietatis of Paracelsus. This noble medicament is of very hot and thin parts, containing all the virtues of the natural balsam, conserving nature in extreme age. It cures quartans, and dissolves the stone; it quickens all the senses, and strengthens the brain and memory; it cures the vertigo, lethargy, epilepsy, headache, convulsions, palsy, pleurisy, jaundice, consumption, catarrh, pestilent fevers, gout, and sciatica; it expels melancholy, and makes the heart glad. Lastly, diseases proceeding from either heat or cold by a certain occult property it strangely cureth. Dose, 6 to 30 grains or more, in wine or other convenient vehicle.

OIL OF SULFUR[s29]
(Oleum Sulfuris[s29] per Campanum)

Rx. According to the old dispensatories it is prepared in a large bell still by the burning and consuming of a large quantity of sulfur[s29] [62], by which a sharp spirit, flying from the kindled sulfur[s29] and beating against the sides of the still, will turn into liquor and flow down like water or oil.

Here it is to be noted:

1. That the sulfur[s29] be put into an earthen cup having sand in it, lest being inflamed it should break it.

2. That this be placed upon another earthen cup, the bottom turned upwards, and these thus disposed be set in the midst of a great earthen pan, then with a red hot iron inflame the sulfur[s29].

3. That these be covered with a great glass bell, or glass funnel, with a neck as long as that of a bolthead, having a hole at top to give

---

[62] Natural mined sulphur, not obtained by steam or Brasch method (Hans Nintzel).

breathing, that the flowers may fly away, by which means you will have a greater quantity and more effectual oil.

4. That it be done in a close, moist place (as in a cellar) and on a moist day.

5. That you leave an empty space between the brims of the bell and the pan, that there may be air enough to keep the sulfur[s29] inflamed.

6. That by reason of the hole at top of the bell or funnel the more phlegmatic part evaporates, while the acid spirits, not being able to rise so high, condense against the sides of the glass.

7. That this spirit is nothing more than a spirit of mined nitre[s1], drawn from a vitriolic salt[s5] in the sulfur[s29].

8. That from 1 pound of sulfur[s29] you will have 1 ounce[s11] of spirit.

It eases all pains of the gout, and that only by bathing (mixing of it with water or SV.). It strengthens the nerves and muscles, and cures a confirmed leprosy. It cures hectics, consumptions, asthmas, and ulcers of the lungs. It makes the teeth white, restores radical moisture, extinguishes all preternatural heat, purifies the blood, and renovates the whole body, expelling putrefaction. In the French-pox it is excellent, and may serve instead of a diet. There is nothing more powerful in expelling of poison, plague, and all pestilent and malign fevers. Dose, 6 to 16 drops or 20 in broth, beer, ale, wine, or cordial julep. Taken alone it kills[63].

ELIXIR OF SUBTILITY

Rx. Of olive oil, honey, SV. rectified and tartarized (Menstruum 21), in equal parts, distil[s16] them all together in ashes, then filter[s30] all

---

[63] Most of these preparations are highly poisonous! Do not try this at home.

the phlegm from the oils, which will be distinguished by the colors; put all of these into a pelican, and add to them a third part of the essence of balm and celandine; digest[s2] it for a month and keep for use.

This is the Elixir Subtilitatis of Paracelsus[64]. It not only resists putrefaction but also preserves all things from putrefaction which appertains to animate bodies. This is the Balsam of Philosophers, which no sensible body is able to resist, it being subtile and able to penetrate everything; it opens all obstructions in the body after a wonderful manner, with many other things, whose virtues are not fitting to be declared only to the Sons of Art.

THE GREAT ESSENCE
(Essentia Magna)

Rx.   Rosemary, Lavender, Sage, Marjoram, Thyme, Balm, Angelica, all full of juice[65]; bruise all in a mortar diligently, pour on a sufficient quantity of malmsey-wine (some say of their own spirit or tincture), then in a vessel with a blind head set it to digest[s2] in balneo with a gentle heat for two months, express all with a press, calcine[s26] the fæces, and extract a crystalline or sweet salt[s5], which add anon (or instead thereof crystals of potassium bitartrate[s21]). This

---

[64] *The Art of Distillation*, p. 45, relates this recipe as follows, "Take Oil Olive, Honey, rectified Spirit of Wine, of each a pint, distil them all together in ashes, then separate all the flegm from the Oils, which will be distinguished by many colours, put all these colours into a Pellican, and adde to them the third part of the Essence of Balm, and Sallendine, digest them for the space of a month. Then keep it for use. This Liquor is so subtile that it preventeth every thing."

[65] Fresh herbs, not dried.

expressed juice or wine digest[s2] for two months, as before, till a glorious liquor be separated from the fæces or sediment, which decant, adding to it the aforesaid crystals and a little oil of cinnamon.

Its virtues are so great that they can scarcely be numbered; for it strengthens all the inward parts, perfectly cures consumptions, all diseases of the head, heart, breast, and lungs, and makes a sad, drooping spirit merry; it cures plague, malign fevers, small-pox, poisons, etc. It is vain to enumerate its virtues (as curing the vertigo, epilepsy, megrim, convulsions, palsy, etc.), but rather advising all to have it by them upon any occasion. Dose: to 1/2 ounce morning, noon, and night, in broth, fragrant wine, or milk.

MAGISTERY OF URINE[s20]

Rx.   Defecated urine[s20], place it in a glass body in BM. for 40 days that it may putrefy; then distil[s16] with a gentle fire in BM. till the phlegm is drawn off; rectify the spirit in a glass with a long wide neck, so have you the volatile salt[s5] which take, and cast away the phlegm; distil[s16] what remains in sand, and a more volatile salt[s5] arises; of the caput mortuum[s18] make a fixed salt[s5], which mix with thrice as much clay, form it into balls, dry them, and distil[s16] by a retort, so have you the spirit of the fixed salt[s5], into which drop the former spirit or volatile salt[s5], till the noise ceases, then sublimate[s28] in sand. So have you a fine, pleasant and delectable salt[s5] of urine[s20], or the true Magistery of salt[s5].

This cuts and dissolves the tartarous coagulum in the whole body, and expels all ill humors; it preserves from the stone, taken once a month before the new moon, and cures consumptions wonderfully. Dose, 10 grains.

MAGISTERY OF SALT[s5]
(According to Mynsicht)

Rx.   Crystalline salt[s5] of wormwood, upon which drop rectified spirit of salt[s5] (Menstruum 8) so much as will coagulate and unite with a prevailing vapor and force above the spirit of salt[s5].

It is a most excellent medicine, having a balsamic property; it renovates the whole man, purifies the blood, strengthens the head, heart and stomach, opens obstructions of both liver and spleen; cuts, discusses, and cleanses from all putrefaction. It is one of the most efficacious diuretics, and a specific against the dropsy, taking away all flatulent, watery, and tartarous viscosity. It breaks and expels the stone; is prevalent against the iliac and cholic passion, the jaundice, all sorts of fevers, palsies, apoplexies, gouts, leprosies, worms, ruptures, etc. Dose, 6 to 16 grains. In a few days it dissolves the dropsy tympanites.

LIQUOR ALKAHEST
(Paracelsi)

Rx.   1. Prepare an alkali from quicksliver[s31] and salt[s5] regale by cementing the salt[s5] regale with the quicksilver[s31], and boiling them in fair water[s4] to make a lixivium, filtrating and coagulating by evaporation; of which salt[s5] prepare a large quantity.

2. Let pure Spanish or Hungarian mercury[s23] be beaten with the alkali, in a stone mortar a little warmed, so long till none of the mercury[s23] can be seen, which put immediately into a glass retort, with a receiver luted well to it, and distil[s16] with a naked fire[s8],

which operation so often repeat till the mercury[s23] becomes very liquid, and appears truly spiritual.

3. This spiritual mercury[s23] distil thrice in a tubulated retort closely joined with *lutum sapientiae*[66] to two other vessels with necks at both ends, the latter of which let be well luted to a large receiver, the mercury[s23] let be cast into the receiver by the tube, which tube ought to rise above the furnace, and after the injection of the mercury[s23] every time to be close stopped, which then pass with its aquosity into its receiver, till it is all turned into water[s4].

4. Put this mercurial[s23] water[s4] into a bolt-head so large as it may fill about an eighth part of it, which for some months place in digestion[s2] in an equable heat, till all the water[s4] is converted into froth. The putrefaction still continues until the froth vanishes, and the liquor in the bottom of the vessel be again clear. At last, rectify it once or twice by a retort in sand, and keep it carefully for use.

Whether this be the Alkahest of Paracelsus with which he did such wonders, and which Helmont so praises, even to the skies, is very doubtful; for as that Alkahest was destinated to the preparation of all sorts of medicines, extraction of all sorts of tinctures and essences, whether out of minerals, vegetables, or animals, so also it was reported to cure effectually all diseases, and to root out the seminations of every malady, and to do miracles above any other medicine except the Philosophers' Stone. He that desires to know more hereof may fetch it out of the most learned Helmont, where he may indeed receive a very great deal of satisfaction. The salt[s5] regale mentioned above is made thus:

---

[66] Lute was commonly used in distillation, which required airtight vessels and connectors to ensure that no vapours were lost; thus it was employed by chemists and alchemists, the latter being known to refer to it as "lutum sapientiae" or the "lute of Wisdom". See bibliographic entry by Stanton Linden.

Rx.   Calcine[s26] the caput mortuum[s18] of the simple spirit, or new potassium bitartrate[s21] (which you please), in a potter's furnace, make a lixivium in water, filter and evaporate; if it be not white enough, dissolve, filter, and evaporate again.

ALL—HEAL OF PARACELSUS
(Panacea Theophrasti Paracelsi)

Rx.   Of the highest rectified spirits of balm, of mugwort, of valerian, of burnet, of juniper, 1 ounce[s11] of each; quintessence[67] of copper[s10], oil of salt[s5] (Menstruum 8), white sugar candy 6 drams[s14], mix them, and keep the mixture in a glass with a glass stopper.

This noble medicine preserves the body, as Hortius says, per totam vitam integrum, in health during the whole life. It takes away heaviness of the head, cures the apoplexy, palsy, epilepsy and other dangerous diseases of the head; it sharpens the eyesight, stays vomiting, and strengthens a weak stomach; it helps asthmas, and most diseases of the lungs; it corrects the vices of the liver and spleen; it is profitable against leprosy, jaundice, colic, stone, disaffections of the womb, and many other diseases. Dose, 1 scruple[s12] to 1/2 dram[s14].

SPIRIT OF FIVE THINGS
(Spiritus Diapente)

Rx.   Paracelsus' Elixir of Property 1 1/2 ounce[s11]. Spiritus theriacalis camphorated, 1 1/2 dram[s14]. Spirit of mined nitre[s1] rectified, 1 dram[s14] (Menstruum 13). Spirit of potassium bitartrate[s21] rectified 2 scruples[s12] (Menstruum 10a). Spirit of salt[s5] rectified

---

[67] QE. in the original text. Quinta essentium.

(Menstruum 8) in which let leaves of gold[s15] No. X. be dissolved[s3] , 1 dram[s14]. Mix all together and digest[s2] twenty days and keep it for use.

This powerfully resists all putrefaction, is an antidote against poison, plague, and small-pox, opens all obstructions of the liver and spleen, purges both reins and bladder, is excellent against the dropsy, and all manner of hot and burning fevers, palsy, jaundice, etc. Dose, 1 scruple[s12], in generous wine.

THE MERCURIAL EAGLE
(Aquila Coelestis)

Rx. Sublimated corrosive (made with salt[s5] and mined nitre[s1]) from which extract a yellow tincture with distilled[s16] vinegar[s22] in ten weeks; decant, and abstract the decanted liquor to dryness, so you have the Aquila Cœlestis in a red powder.

This medicine is commend by Paracelsus almost in the highest degree for curing the French-pox, gout, epilepsy, and most diseases of the head, rooting them out by sweat. Dose, 2 to 4 drops.

ELIXIR OF MINED NITRE[s1]
(According to Mynsicht)

Rx. Of galanga the less 1 1/2 ounce[s11], calamus aromaticus 1 ounce[s11], mint, red sage, each 1/2 ounce[s11], choice cinnamon, cloves, ginger, in equal parts 3 drams[s14], nutmegs, cubebs, 2 drams[s14] each, xyloaloes, citron peels, 1 dram[s14] each. Mix and make a powder; add white sugar candy 3 ounces[s11] SV. rectified so much as to make it thick like honey. Put all into a glass, and put thereto oil[s7] of the mined nitre[s1] of copper[s10] or iron[s24], or spirit of mined nitre[s1] often rectified, so much as to overtop it the breadth of four fingers. Digest[s2] 40 days, at length decant the tincture and filtrate; upon the remaining fæces put SV., and according to the spagyric art extract an essence; mix both these together, circulate in BM. for 20 days, and keep it for use.

Experience testifies that there is scarcely a more noble and efficacious stomachic in the whole republic of medicine. It is a great secret in all affects of the ventricle, given in mint-water, for it comforts all the inward parts and principal members; it cools heats, and causes appetite; it is most excellent in the epilepsy, apoplexy, catarrhs, phlegmatic disposition of the whole body, pain of the head, lethargy, and fevers. Dose, 1/2 scruple[s12] to 1 scruple[s12] in appropriate liquor.

PHILOSOPHIC SPIRIT OF SALT

Rx. Salt[s5] as much as you like, suppose 8 ounces[s11], oil of mined nitre[s1] rectified 4 ounces[s11], water[s4] q.s. for filtration[s30] of the salt[s5]. First there comes off a phlegm, then put the matter into a glass retort and distil[s16] in sand, then change the receiver, and distil[s16] to dryness; so will you have the most exalted spirit of salt[s5]. But the oil[s7] of mined nitre[s1] will be coagulated in the bottom of the retort with the alkali or caput mortuum[s18] of the common salt[s5]. Take of this acid spirit 2 ounces[s11] salt[s5] (dissolved in water[s4]) 1 ounce[s11] mix and distil[s16] in sand, as before; so will you have 3 ounces[s11] of spirit. Thus may you proceed infinitely, and increase the quantity of spirit with little or no charge as long as you please.

This spirit has many uses in chemistry for dissolving bodies, precipitating of things dissolved in aqua fortis, spirit of vinegar, etc., and in extracting tinctures. Inwardly taken in wine, ale, or water, it opens, cools, resists putrefaction, takes away all manner of fevers and unnatural heats, and is a potent remedy against the plague. It strengthens the stomach, fortifies the heart, cheers the spirits, and refreshes wearied and decayed nature. Dose from 10 drops to 20 in any proper liquor or vehicle.

OIL OF GOLD[s15]
(Oleum Solis vel Auri)

Rx. Quicklime[s25] of gold[s15] made by reverberation with royal cement; cleanse it and digest[s2] it 24 hours in rectified Aqua Vitae, so shall you have oil[s7] of gold[s15]. Thus Paracelsus. Or thus:

Rx. Of the sharpest juice of lemons filtrated 6 ounces[s11], leaves of fine gold[s15]. No. 60, digest[s2] them in a glass vessel with a gentle heat for four or five days, then filter and abstract the juice by distillation[s16] and the gold[s15] will remain in the bottom in the form of butter. Thus Gesner[68].

This mixed with wine will give it the color of gold. It wonderfully resists putrefaction; it also purges, and moves to sweat; it cures the leprosy, and such as has been spoiled by mercurial unguents.

NOTE: The above reverberation may be done by putting the filings of gold[s15] alone into a crucible in a reverberatory furnace and burning them (without melting) till they come to a calx of a purple color, thin and light; or you may mix with the filings or flowers of sulfur[s29], and then reverberate till the calx becomes as aforesaid. Furthermore, often dissolution in aqua regia does the work as well.

---

[68] Conrad Gesner was an alchemist of the 16th century.

THE SILVER HELL-STONE

(Lapis Infernalis Argenteus)

Rx. Filings of fine silver[s9], one part good aqua fortis or spirit of nitre[s6], 2 parts, filter[s30] in a small matrass with a long neck luted half way; evaporate the humidity in a circulary fire[s8] to dryness, leaving a blackish scum on it; then give a melting heat till the fumes cease; take off the matrass, and forthwith cast it into little brass or iron moulds.

It is caustic, remaining forever if kept from air. It consumes by touching warts, proud and dead flesh, cancers, ulcers, etc., if you wet them with a little water; it dyes hair and skin an unchangeable black.

MAGISTERY OF IRON[s24]

(According to Salmon)

Rx. Filter[s30] filings of steel in purified juice of lemons, digest[s2] for a month, then filter into a glass vessel, and in a sand heat inspissate to the consistence of a liquid extract. The remainder of the chalybs which will not go through the paper, dry and reduce into a subtile powder for the same use, or for steeled wine.

Both the liquid extract and the powder have a wonderful force in opening all obstructions, and dissolving all tartareous and coagulated matter, and strengthening all the internal viscera. They are an excellent cure for melancholy, quartans, dropsy, and all diseases of the womb occasioned through obstruction. Dose of the extract 1/2 ounce[s11] to 1 dram[s14]; of the powder, 1 to 2 scruples[s12].

MAGISTERY OF IRON[s24] (Vitriolated)

Rx.   Dissolved steel in rectified spirits of mined nitre[s1], then coagulate, so have you a magistery green like vitriol.

It opens obstructions of the liver and spleen, and cures the jaundice, quartans, melancholy, and the green sickness. Dose, 1/2 ounce[s11] in Rhenish wine.

WATER AND OIL OF POTASSIUM BITARTRATE[s21]

Rx.   Mined nitre[s1] calcined[s26] to whiteness 2 pound, white potassium bitartrate[s21] 1 pound, powder and mix them, then distil[s16] by a retort; the water separated from the oil rectify.

The rectified water, mixed with a sufficient quantity of rose-water, and dropped in the eyes[69] , cures most distempers happening to them. Taken inwardly in Rhenish wine, it opens obstructions and cures the green sickness.

ELECTRUM, OR GOLDEN ELIXIR OF ANTIMONY[s19]

Rx.   Regulus of antimony[s19] (made of Mercurius Vitæ, and as much potassium bitartrate[s21] and nitre[s6]) fine gold[s15] , 1/2 ounce[s11], melt and powder them together, to which put ammonium chloride[s13] 2 ounces[s11]; sublimate[s28] till the star of antimony[s19] ascends, and a useless earth remains at bottom; wash off the salt[s5], and put the remaining gold[s15] and flowers of antimony[s19] in the bottom into an Hermetical egg, in a fixed capella, with such a fire[s8]

---

[69] Do not try this at home!

as may not melt them, but may make them of a yellow citron color, and afterward of a chestnut color. Extract these flowers with spirit of vinegar[s22] alkalized, and then draw out the tincture with SV. as before taught.

It cures consumptions, hypochondriac melancholy, black jaundice, dropsy, gout, and scurvy. Dose, the quantity of 1 grain twice a day in some proper syrup.

WATER AGAINST CANKERS
(Aqua Phagedoenica)

Rx. Make a strong lixivium of quicksilver[s31] in boiling water[s4], filter it into a glass bell, in 1 pound of which dissolve[s3] corrosive sublimate 1/2 dram[s14] , stir them and there will be an orange color, and the sublimate will fall to the bottom. If the water[s4] be too strong put upon it more lime water[s4] till it is as you would have it. But Fallopius makes it by putting into lime water[s4] 1 pound Mercurius Dulcis a sufficient quantity (viz. 2 ounces[s11]) and dissolving it by boiling.

Either of these Phadegaenick[70] waters are good against inflammations, fistulas, malignant and venereal ulcers, cankers, scabs, sores, pustules, and other breakings-out, as also the itch, leprosy and the like, curing them without danger; first wash well with the water, then apply to the affection a linen cloth dipped therein, so will the proud flesh (if there be any) be consumed, the putrefaction corrected, the sore cleansed and incarnated, and at last by its drying quality cicatrized[71] and made well.

---

[70] *Phadegaenick* might refer to a curative water for phagedenic ulcers.
[71] Healed by forming scar tissue.

John Hazelrigg

Symbol Dictionary

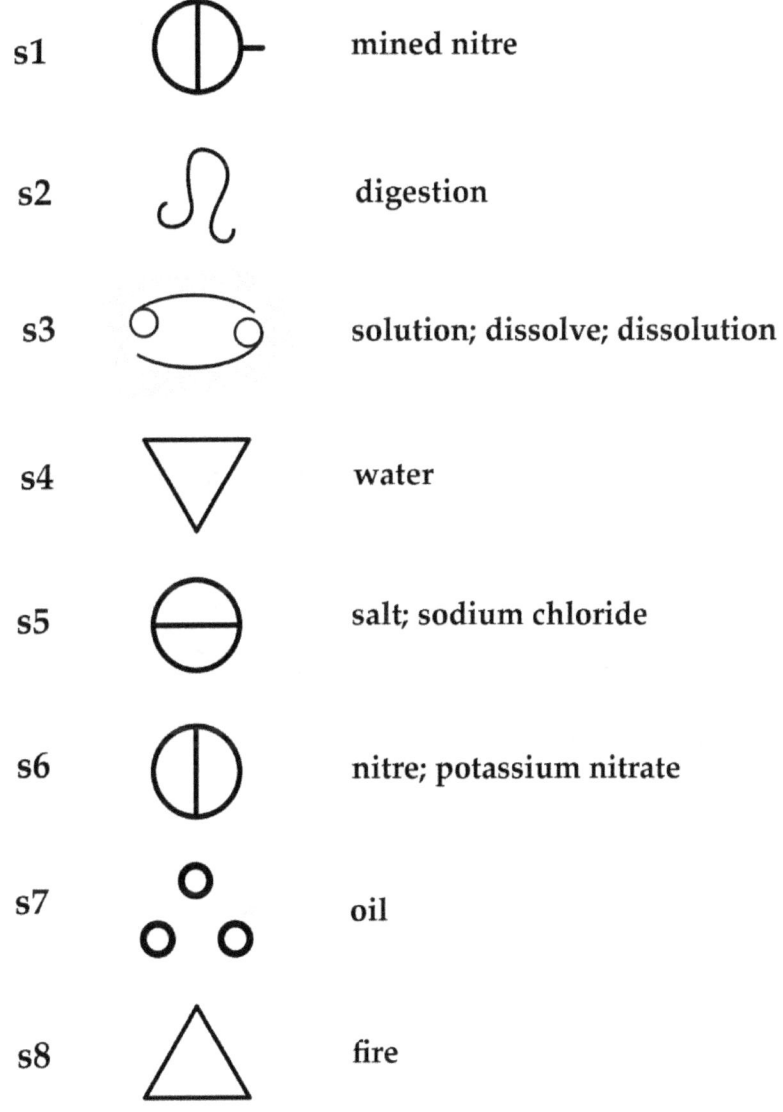

s1   mined nitre

s2   digestion

s3   solution; dissolve; dissolution

s4   water

s5   salt; sodium chloride

s6   nitre; potassium nitrate

s7   oil

s8   fire

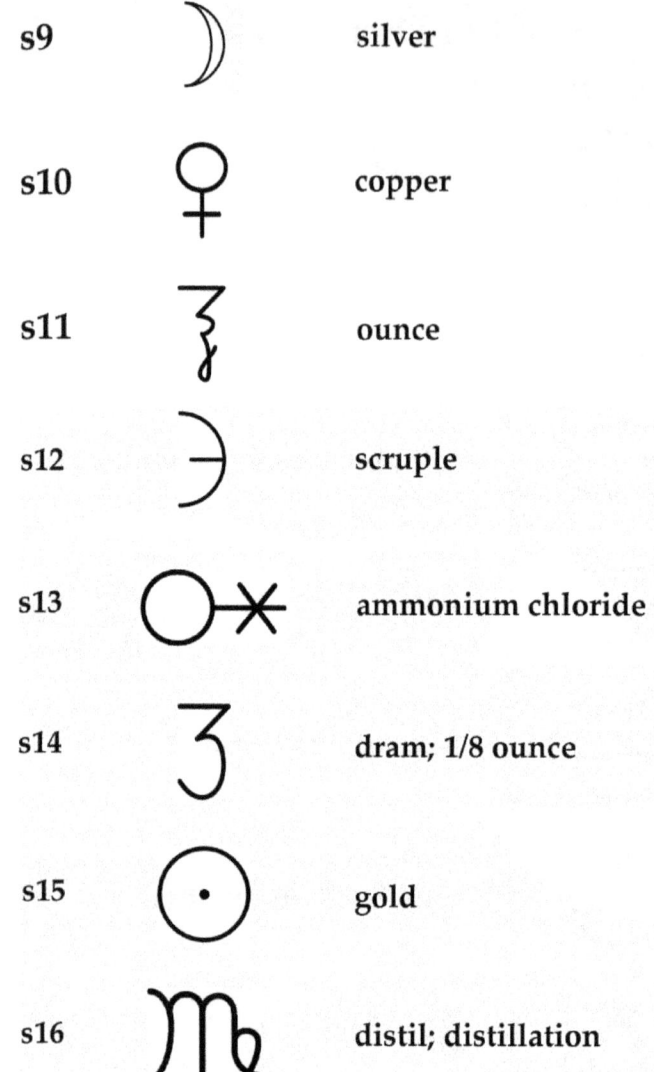

s9	silver
s10	copper
s11	ounce
s12	scruple
s13	ammonium chloride
s14	dram; 1/8 ounce
s15	gold
s16	distil; distillation

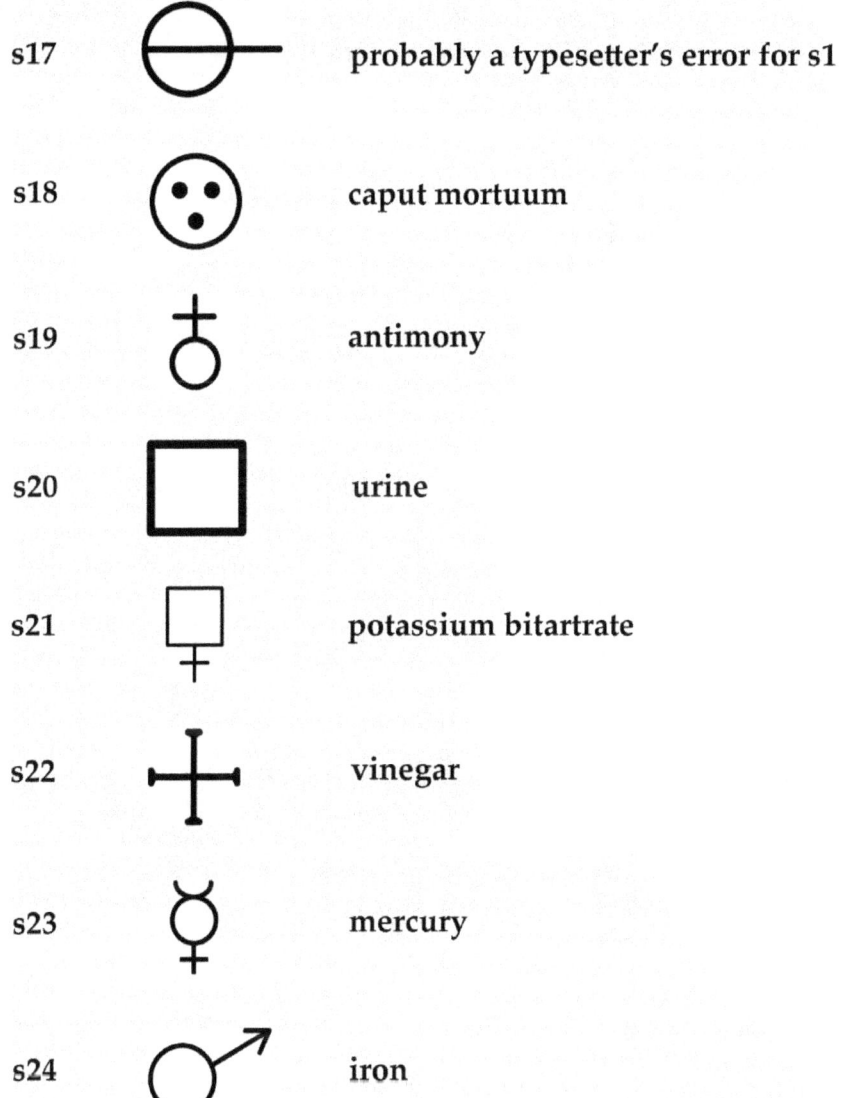

s17   probably a typesetter's error for s1

s18   caput mortuum

s19   antimony

s20   urine

s21   potassium bitartrate

s22   vinegar

s23   mercury

s24   iron

s25	Ψ	quicklime
s26		calcine; calcinated
s27		probably a typesetter's error for s29
s28		sublimate; sublimation
s29		sulphur
s30		filter; filtration
s31		quicksilver

BIBLIOGRAPHY

French, John. *Art of Distillation, The*. London: Richard Cotes, 1651.

Hazelrigg, John. *Book of Formulas, The*. New York: Hermetic Publishing Co., 1904.

Hazelrigg, John. *Book of Formulas, The*. Richardson TX: Restorers of Alchemical Manuscripts, 1977.

Linden, Stanton J. *Alchemy Reader: from Hermes Trismegistus to Isaac Newton, The*. Cambridge: Cambridge University Press, 2003.

Wheeler, Philip N. *Tables of Alchemical Symbols and Gematria*. Waynesboro VA: Alembic Publishing, 2011.

A Word from the Publisher

Thank you for purchasing this book!

The folks at Alembic Publishing hope that you enjoy this small work. Our vision is to reproduce the best quality works by ancient and modern alchemists.
Please visit our web site for a complete list of our available titles:

http://www.AlembicPublishing.com

Ora et labora!

Alembic Publishing

www.ingramcontent.com/pod-product-compliance
Lightning Source LLC
Chambersburg PA
CBHW071610170526
45166CB00003B/1039